Pedestrian Safety in Sweden

PUBLICATION NO. FHWA-RD-99-091

DECEMBER 1999

U.S. Department of Transportation
Federal Highway Administration

Research, Development, and Technology
Turner-Fairbank Highway Research Center
6300 Georgetown Pike
McLean, VA 22101-2296

FOREWORD

Creating improved safety and access for pedestrians requires providing safe places for people to walk, as well as implementing traffic control and design measures which allow for safer street crossings. A study entitled "Evaluation of Pedestrian Facilities" involved evaluating various types of pedestrian facilities and traffic control devices, including pedestrian crossing signs, marked versus unmarked crosswalks, countdown pedestrian signals, illuminated pushbuttons, automatic pedestrian detectors, and traffic calming devices such as curb extensions and raised crosswalks. The study provided recommendations for adding sidewalks to new and existing streets and for using marked crosswalks for uncontrolled locations. The "Evaluation of Pedestrian Facilities" also included synthesis reports of both domestic and international pedestrian safety research. There are five international pedestrian safety synthesis reports; this document compiles the most relevant research from Scandinavia.

This synthesis report should be of interest to State and local pedestrian and bicycle coordinators, transportation engineers, planners, and researchers involved in the safety and design of pedestrian facilities within the highway environment.

Michael F. Trentacoste
Director, Office of Safety
Research and Development

NOTICE

This document is disseminated under the sponsorship of the Department of Transportation in the interest of information exchange. The U.S. Government assumes no liability for its contents or use thereof. This report does not constitute a standard, specification, or regulation.

The U.S. Government does not endorse products or manufacturers. Trade and manufacturer's names appear in this report only because they are considered essential to the object of the document.

FOREWORD

The aim of this report is to highlight some of the more important research findings in Scandinavia. Since our position is Lund in Sweden, most of our findings derives from Swedish results. The Scandinavian countries have great similarities and most of the findings are valid for all the countries. In some cases the difference in the traffic law plays an important role. This is especially the case for zebra crossings where the rules in Norway are significantly different from the other Scandinavian countries.

Technical Report Documentation Page

1. Report No. FHWA-RD-99-091	2. Government Accession No.	3. Recipient's Catalog No.
4. Title and Subtitle Pedestrian Safety in Sweden		5. Report Date
		6. Performing Organization Code
7. Author(s) Dr. Lars Ekman & Dr. Christer Hyden		8. Performing Organization Report No.
9. Performing Organization Name and Address Lund University Dept. of Traffic Planning & Engineering University of North Carolina Highway Safety Research Center 730 Airport Rd, CB #3430 Chapel Hill, NC 27599-3430		10. Work Unit No. (TRAIS)
		11. Contract or Grant No. DTFH61-92-C-00138
12. Sponsoring Agency Name and Address Federal Highway Administration Turner-Fairbank Highway Research Center 6300 Georgetown Pike McLean, VA 22101-2296		13. Type of Report and Period Covered
		14. Sponsoring Agency Code

15. Supplementary Notes

Prime Contractor: University of North Carolina Highway Safety Research Center
FHWA COTR: Carol Tan Esse

16. Abstract

This report was one in a series of pedestrian safety synthesis reports prepared for the Federal Highway Administration (FHWA) to document pedestrian safety in other countries. Reports are also available for:

United Kingdom (FHWA-RD-99-089)
Canada (FHWA-RD-99-090)
Netherlands (FHWA-RD-99-092)
Australia (FHWA-RD-99-093)

This report is a review of recent pedestrian safety research in Sweden (in particular) with some attention to similar research in other Scandinavian countries. The report states that even in Sweden, where attention has long been paid to pedestrian and bicyclists concerns, even so, still too much traffic planning is addressed as if it were a vehicular issue only.

If traffic cannot be separated, then consideration should be given in some areas to restricting vehicle speeds to 30 km/hr. It is argued that future planning must better balance the competing needs of motor vehicle traffic, pedestrians, and cyclists.

17. Key Words: Pedestrians, safety, Sweden, walking cycling		18. Distribution Statement		
19. Security Classif. (of this report) Unclassified	20. Security Class f. (of this page) Unclassified		21. No. of Pages 37	22. Price

Form DOT F 1700.7 (8-72) Reproduction of form and completed page is authorized

SI* (MODERN METRIC) CONVERSION FACTORS

APPROXIMATE CONVERSIONS TO SI UNITS

Symbol	When You Know	Multiply by	To Find	Symbol
LENGTH				
in	inches	25.4	millimeters	mm
ft	feet	0.305	meters	m
yd	yards	0.914	meters	m
mi	miles	1.61	kilometers	km
AREA				
in^2	square inches	645.2	square millimeters	mm^2
ft^2	square feet	0.093	square meters	m^2
yd^2	square yards	0.836	square meters	m^2
ac	acres	0.405	hectares	ha
mi^2	square miles	2.59	square kilometers	km^2
VOLUME				
fl oz	fluid ounces	29.57	milliliters	mL
gal	gallons	3.785	liters	L
ft^3	cubic feet	0.028	cubic meters	m^3
yd^3	cubic yards	0.765	cubic meters	m^3

NOTE: Volumes greater than 1000 l shall be shown in m^3.

Symbol	When You Know	Multiply by	To Find	Symbol
MASS				
oz	ounces	28.35	grams	g
lb	pounds	0.454	kilograms	kg
T	short tons (2000 lb)	0.907	megagrams (or "metric ton")	Mg (or "t")
TEMPERATURE				
°F	Fahrenheit temperature	5(F-32)/9 or (F-32)/1.8	Celcius temperature	°C
ILLUMINATION				
fc	foot-candles	10.76	lux	lx
fl	foot-Lamberts	3.426	candela/m^2	cd/m^2
FORCE and PRESSURE or STRESS				
lbf	poundforce	4.45	newtons	N
lbf/in^2	poundforce per square inch	6.89	kilopascals	kPa

APPROXIMATE CONVERSIONS FROM SI UNITS

Symbol	When You Know	Multiply by	To Find	Symbol
LENGTH				
mm	millimeters	0.039	inches	in
m	meters	3.28	feet	ft
m	meters	1.09	yards	yd
km	kilometers	0.621	miles	mi
AREA				
mm^2	square millimeters	0.0016	square inches	in^2
m^2	square meters	10.764	square feet	ft^2
m^2	square meters	1.195	square yards	yd^2
ha	hectares	2.47	acres	ac
km^2	square kilometers	0.386	square miles	mi^2
VOLUME				
mL	milliliters	0.034	fluid ounces	fl oz
L	liters	0.264	gallons	gal
m^3	cubic meters	35.71	cubic feet	ft^3
m^3	cubic meters	1.307	cubic yards	yd^3
MASS				
g	grams	0.035	ounces	oz
kg	kilograms	2.202	pounds	lb
Mg (or "t")	megagrams (or "metric ton")	1.103	short tons (2000 lb)	T
TEMPERATURE				
°C	Celcius temperature	1.8C+32	Fahrenheit temperature	°F
ILLUMINATION				
lx	lux	0.0929	foot-candles	fc
cd/m^2	candela/m^2	0.2919	foot-Lamberts	fl
FORCE and PRESSURE or STRESS				
N	newtons	0.225	poundforce	lbf
kPa	kilopascals	0.145	poundforce per square inch	lbf/in^2

*SI is the symbol for the International System of Units. Appropriate rounding should be made to comply with Section 4 of ASTM E380.

(Revised September 1993)

TABLE OF CONTENTS

1. Pedestrian safety situation in Scandinavia 1
   1.1 Result from police reported accidents 1
   1.2. Result from hospital records 2
       Most pedestrians are injured in single accidents 2
   1.3. Other sources for safety information 2
       The traffic conflict technique 2

2. The effect of common pedestrian facilities 4
   2.1. Zebra crossings 4
       With and without refuge 10
   2.2. Small roundabouts 12
   2.3. Traffic calming 14
       Environmentally adapted through-roads 14
   2.4 Project WALCYNG — How to encourage WALking and CYcliNG instead of
       shorter car trips and to make these modes safer 14
       Introduction 14
       Main findings 15
       a. What is known about the target groups and their situation? 15
       b. What is known about the preconditions for WALCYNG? 16
       The WALCYNG Quality Scheme 19
   2.5 Car - Pedestrian interaction at zebra crossings 19

3. The use of new pedestrian facilities 21
   3.1. Detection of pedestrians at signal-controlled intersections 21
   3.2. Relevant warning system 22
   3.3. Warning lights mounted at the roadways 23
   3.4. Painted pre-marking at zebra crossings in Stockholm 24
   3.5 Fluorescent caps on first class pupils 25
   3.6 Ultra violet light (UV-Light) 26

4. Ongoing and future research in Scandinavia 26
   4.1. The implementation of the Vision Zero 26
       A new approach to road safety 27
       System designer has primary responsibility 27
       Action in a variety of fields is needed to produce a safe road system 28
   4.2. Speed Limiters for controlling vehicle speeds 28
       Interview results 29
       Driving behaviour 31
       Driving patterns 31
       Speed profiles 31

5. Conclusion 32

6. References 33

1. Pedestrian safety situation in Scandinavia

1.1. Result from police reported accidents

Police reported accidents are still the most common source of information if we want to study the overall safety situation for road users. In the Scandinavian countries, we have a long tradition of studying accidents reported by the police. Since the accuracy of these reports decline with declining severity, we tend to only rely 100 percent on fatal accidents; but for most purposes, all injury accidents are used. Looking at the number of people killed in road related accidents, we can see that pedestrian accidents still constitutes a severe problem.

Table 1. Police reported pedestrian accidents 1995 in the Scandinavian countries.

	Population (10^6)	Number of killed pedestrians	Number of injured pedestrians	Killed and injured	Number of killed pedestrians per million inhabitant
Sweden	8.8	71	1403	1474	8.1
Denmark	5.2	118	1033	1151	22.7
Finland	5.2	72	1031	1103	13.8
Norway	4.3	51	1105	1156	11.9

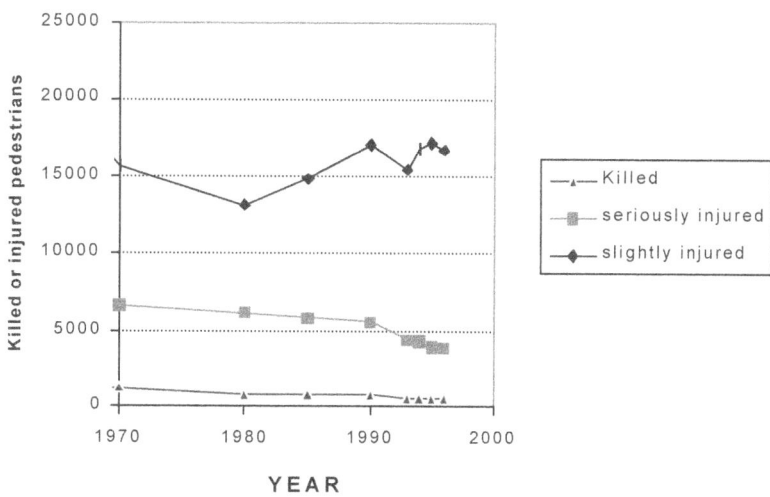

Figure 1. Police reported accidents in Sweden, 1970-1996.

1.2. Result from hospital records

Since the problem with underreporting is especially big for pedestrian and bicycle accidents, it makes sense to include results from hospital records. In Sweden all accidents on roads or sidewalks — including at least one vehicle — are considered traffic accidents. This includes single accidents for pedal cyclists in the same way as single accidents for other road users, but it does not include single accidents for pedestrians. Besides, single accidents for cyclists are almost non-existing in the police records.

In a study by Berntman and Modén (1996) it was found that more than 400 pedestrians were injured during 1 year in the city of Malmö. This is about five to seven times the number reported by the police. More than half of the injured people are above 65 years old. This is also true for smaller towns. Problems for pedestrians generally increase with increasing age. As an example, we can often find traffic accidents involving pedestrians above 90 in the hospital statistics. Women are more at risk than men. In smaller towns, such as Lund (100,000 inhabitants), 70 people per 100,000 inhabitants are injured in traffic. In Malmö which is a bigger town (225,000 inhabitants) 185 people per 100,000 inhabitants are injured.

Most pedestrians are injured in single accidents

In studies based on hospital records, the most predominant type of accidents involve pedestrians who have fallen or slipped on icy or snowy surfaces. Berntman and Modén (1996) found that between 65 and 80 percent of the accidents reported by the hospital were of this type. Compared to other road user groups, falling and slipping is far more common for pedestrians. In our Nordic climate, icy and snowy sidewalks play an important role in many pedestrian accidents. In a Swedish study (LTH/VTI 1996), it is concluded that the climate and thus snow and ice conditions have a great impact on pedestrian problems. In the city of Göteborg and Lidköping, icy or snowy conditions were present in 50 percent of all accidents in which pedestrians fell. In Umeå, in the far north with colder climate, the number is 65 percent.

Pedestrians injured by falling often claim that bad winter maintenance was a main contributing factor to the accident. Holes and generally uneven surfaces are also often mentioned.

1.3. Other sources for safety information

The traffic conflict technique

From a methodological point of view, the monitoring and evaluation of traffic safety is a very challenging task. Generally, safety monitoring and evaluation must include a number

of different tools. Accident reports normally provide only limited information about the accident generating processes. Besides, it is often difficult to make reliable estimates of the expected number of accidents. Accident analysis must therefore be supported by supplementary information.

The most common concept in this respect is the traffic conflict. It represents a link between behavior and accidents. It has 'one leg' on the behavioral side, via the continuous monitoring of behaviors and specification of behaviors that lead to serious conflicts. The 'other leg' is on the accident side, thanks to the identification of near-accident situations (= serious conflicts) that have shown a close relationship with accidents.

Accident analysis plays the role of large-scale monitoring of safety problems, while conflict studies are used to identify the kind of problems that lie behind the accidents and what kind of measures might be effective. Behavioral studies, finally, support conflict studies through more comprehensive and detailed studies to explain the presence of certain accident- producing behaviors, or why this behavior is more common than other nonaccident producing behaviors.

The Swedish Traffic Conflicts Technique (TCT), the most commonly used today, is based on two concepts: Time to Accident (TA) and Conflicting Speed (CS) (Hydén, 1987). TA is the time that remains from the moment one of the road users takes evasive action until a collision would have occurred if the speeds and directions of the involved road users had not changed. CS is the speed of the road user that takes evasive action, just prior to the evasive action. A serious conflict is defined by certain border values for TA and CS. The serious conflicts are recorded by human observers. The training of observers normally takes 1 week.

One great advantage with conflict studies is that they are easy to perform. In the most simple form, no sophisticated hardware is needed, only a pencil and a recording sheet. Today, image processing is becoming a valid concept for making conflict studies more efficient and reliable.

The Swedish technique has been carefully assessed regarding reliability and validity. It has been demonstrated that serious conflicts very often are more reliable in predicting average expected accident numbers at individual locations than occurred accidents are. Moreover, events leading to accident situations (i.e., to serious conflicts) are more reliably recorded by trained observers than if they were to be recorded via interviews with accident involved road users, or via some other reconstructing procedure.

Long experience in the development and use of this tool has clearly shown that the Swedish TCT is reliable and valid enough to be a useful not only in Sweden but also in other countries. The Swedish technique has been applied in quite a few countries around the world.

Thanks to this comprehensive knowledge of what behaviors actually produce risks and what the underlying reasons for these behaviors are, one can then hypothesize about remedial measures in any of the traditional areas, e.g., road design, campaigns, information, education, etc., thus treating both general problems and local problems. There is one more major point; conflict studies can be used in an after situation to see whether the hypotheses that were formulated actually proved to be true or not. Thus, the learning process, and the increase of useful knowledge, can be accelerated.

The main problem in using the technique today is that it is too demanding of time and human resources. One way out of this problem is to make the procedure of data collection and analysis more or less automatic. Research efforts are underway to link conflict studies to video- recording and image processing.

2. The effect of common pedestrian facilities

2.1. Zebra crossings

Zebra crossings are one of the most common pedestrian safety countermeasures in Sweden, as in all Scandinavian countries. To test the safety effect of a standard zebra crossing, a comparative study was carried out with data collected from five cities in southern Sweden. The vast majority of intersections in central parts of Swedish urban areas have been equipped with marked zebra crossings.

It has been considered almost self evident to most people that it is safer to cross the road at a zebra crossing. The well-known fact that most of the accidents where pedestrians have been hit by a car in urban areas occur either at a zebra crossing or at a signalized intersection has not changed the opinion about the zebra crossing. It was considered obvious that this was caused by the high exposure at zebra crossings and signalized intersections respectively. No one had, however, up to that point, measured the exposure of pedestrians in Sweden (Ekman, 1988).

The study was designed to ensure comparison with previous British studies by Jacobs and Wilsson (1967). In that study only a few streets were examined, but in this study, data were collected from many streets in the five cities. In line with the British study, the street was divided into segments that could be considered comparable.

In addition to the zones defined in the British study, some zones were added to make it possible to compare the zebra crossing, including the area just surrounding the zebra marking, to approaches with no zebra marking or signalized facilities. The reason for this was that the previous studies pointed out that there was a big difference in accident rates between the zebra and the area just beside the crossing itself. This had been used as an indication of the safety benefits of the zebra crossing. The high accident rate close to the zebra crossing should be looked upon as a negative side effect of the zebra.

Table 2. Cities where the data was gathered.

City	Inhabitants in urban areas of the city in 1980
Göteborg	457000
Malmö	227000
Lund	55000
Landskrona	27000
Eslöv	14000

Accident data were collected from the cities listed in table 2 in southern Sweden. Data on police reported injury accidents where pedestrians had been hit by a vehicle was collected from 1979 to 1984. The length of the period was a compromise between a reasonably large data set on one hand and on the other hand avoiding too large changes in the traffic situation during the time period.

Each accident was classified in line with the zone system, which means that only accidents that involved both crossing pedestrians and cars were considered. Accordingly, the number of crossing pedestrians was counted at all streets. To cover all the selected streets, the counting was carried out with a sample technique. All the counting was done manually by observers. In total, 56,700 km (35,211 mi) of street were covered. For each individual zone, the number of crossing pedestrians was counted twice during a short time (two 6-minute periods). The peak traffic period was avoided by counting crossing pedestrians between 9:00 to 12:00 and between 13:00 to 16:00. An analysis of accident occurrence suggested counting the traffic during off-peak time, since the accidents seemed to be dispersed during the day with some concentration in the afternoon.

The accident rate was defined in accordance with the probability approach. The aim was to see if it was safer for those pedestrian crossing the street at a zebra crossing compared to those crossing at other similar locations. The accident rate was then:

$$R = \frac{\text{the number of police reported accidents during 6 years}}{\text{the number of crossing pedestrians during 12 minutes}} * 1000$$

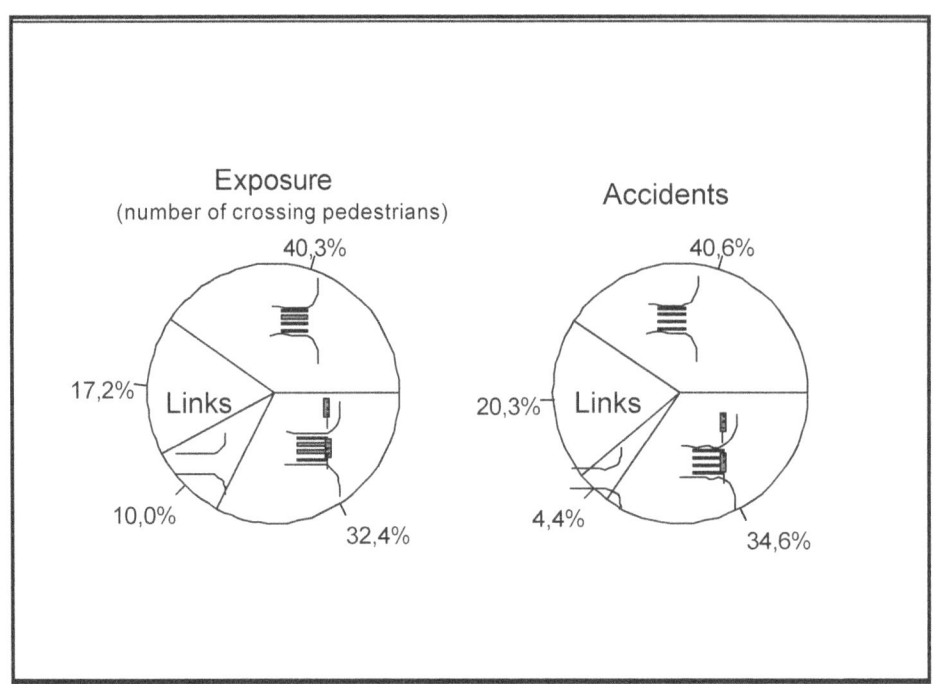

Figure 2. The distribution of exposure and accidents.

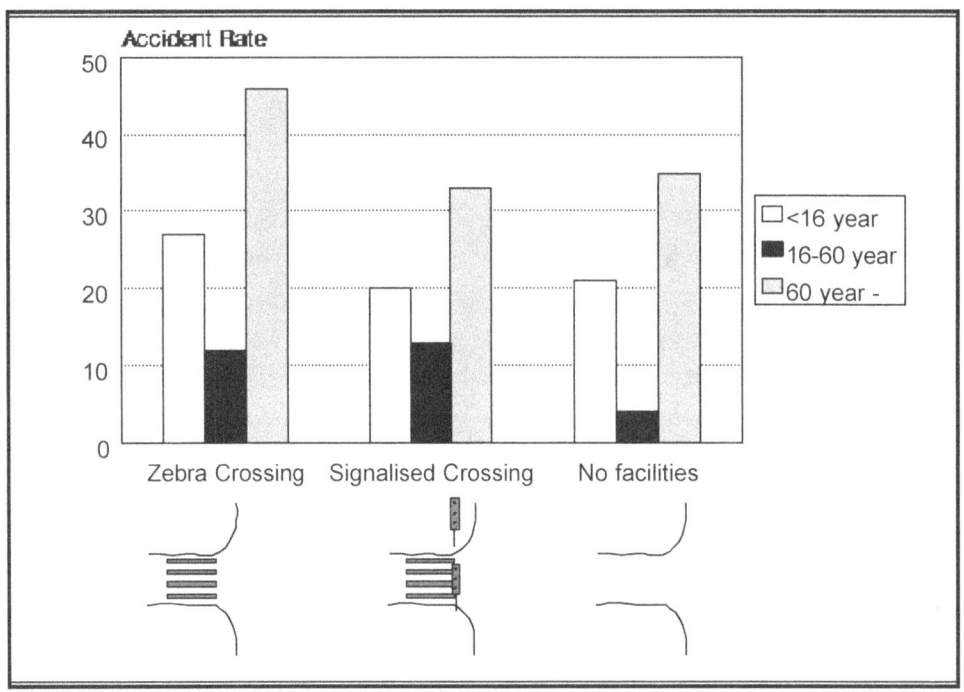

Figure 3. Accident rates for the three crossing types by age groups.

The multiplication by 1,000 was made to get manageable numbers.

The main result was that crossing at intersections where there is zebra marking seems to result in higher risk for an individual pedestrian than crossing at other intersections. It was also shown that signalized intersections do not provide a safe crossing situation for pedestrians. For children and elderly, there seems, however, to be an indication that the signal could slightly improve the situation.

When the weak groups (children and elderly) were excluded, it became obvious that the high accident rates for zebra crossings could not be caused by high number of weak road-users, see figure 3.

The results were not only checked for age effect but also for car flow. The crossing point studied was along long streets with most of the types of zones covered to control car flow. If a street with remarkably high car flow was picked, this would then affect all the zones. As a double check, car flow was compared, and no major difference was found between the types of zones. Signalized intersections and intersections with no facilities were even found to have slightly higher car flow than those locations with zebra crossings.

The general explanation to these remarkable results was that pedestrians experience a false feeling of safety when protected by zebra marking or signalization. Another way of expressing it could be that pedestrians cross more carefully when no help is provided. This study does not cover the mobility gains, if there are any, from zebra crossings or signalization.

Three main conclusions could be drawn from this study:

- The safety potential at signalized intersections is not fully achieved.

- Behavior adaptation or behavior modification is the key to safety improvements or failure.

- The safety potential is great at both zebra crossings and at signalized intersections, since two thirds of all pedestrians cross at these locations.

In a more elaborate study regarding the relationship between accidents or conflicts and exposure, Ekman (1996) compared intersections with and without zebra crossings. The results are in many ways comparable; both studies focus on major streets in partly the same cities. In the latter study, data were collected from the cities of Malmö and Lund. The results are illustrated in the following two graphs.

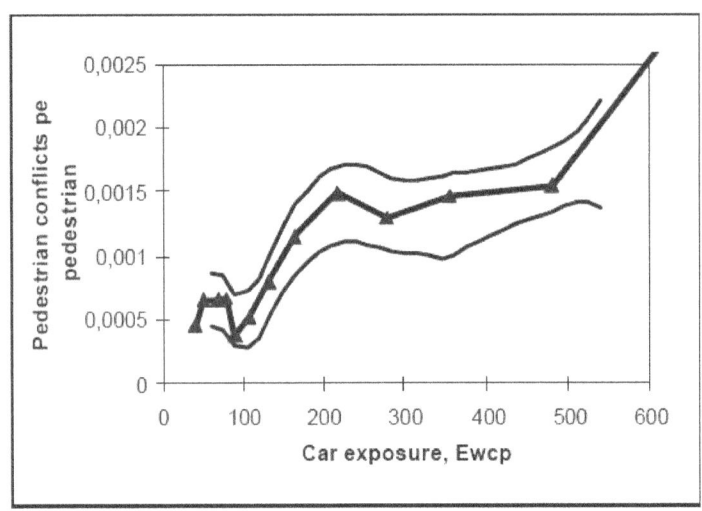

Figure 4. Pedestrian conflicts per pedestrian versus car exposure for approaches with zebra crossing. (Moving average line (RPF[1]) with an estimated 80% confidence interval.)

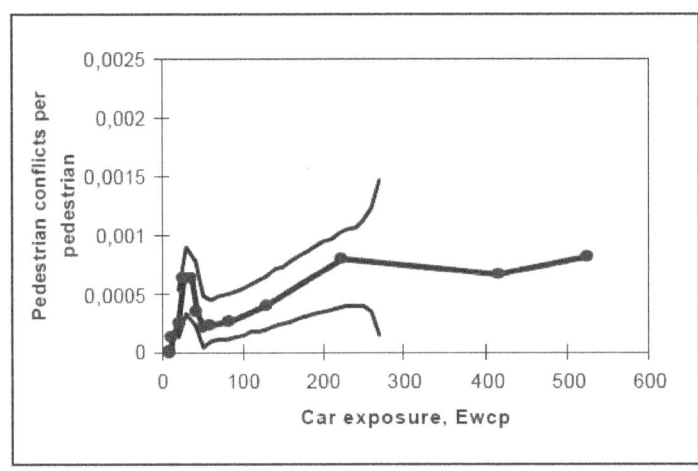

Figure 5. Pedestrian conflicts per pedestrian versus car exposure for approaches without zebra crossing. (Moving average line (RPF[1]) with an estimated 80% confidence interval.)

Two main conclusions could be drawn from the figures above.

- Zebra crossings seem to have higher accident rate than approaches without zebra marking.

---

[1]Risk Performance Functions (RPF).

- The increased accident rate for approaches with zebra crossings is only valid on locations where the car flow is larger than 100 cars per hour.

This data set does not include enough locations without zebra crossings and with car exposure higher than 250 cars per hour to allow the construction of a valid confidence interval.

The locations, with and without zebra crossings, do not only vary according to the car exposure. The pedestrian flow is naturally much higher at locations with a zebra crossing. If we look at the RPF of pedestrian flow we get the results presented in figures 6 and 7.

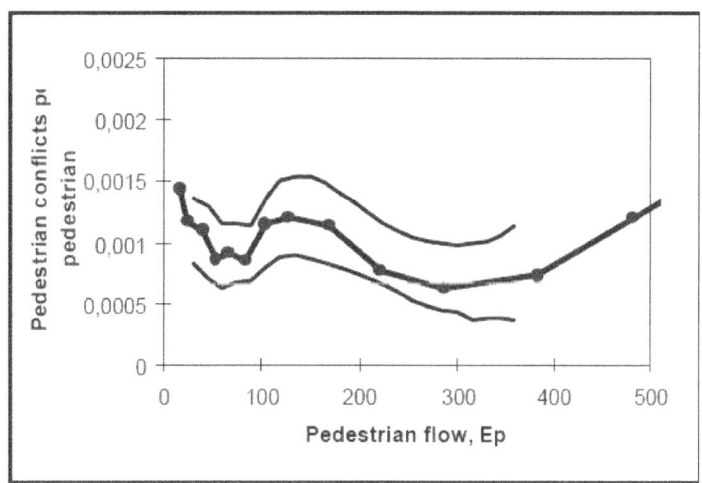

Figure 6. Pedestrian conflicts per pedestrian versus pedestrian flow for approaches with Zebra crossing. (Moving average line (RPF) with an estimated 80% confidence interval.)

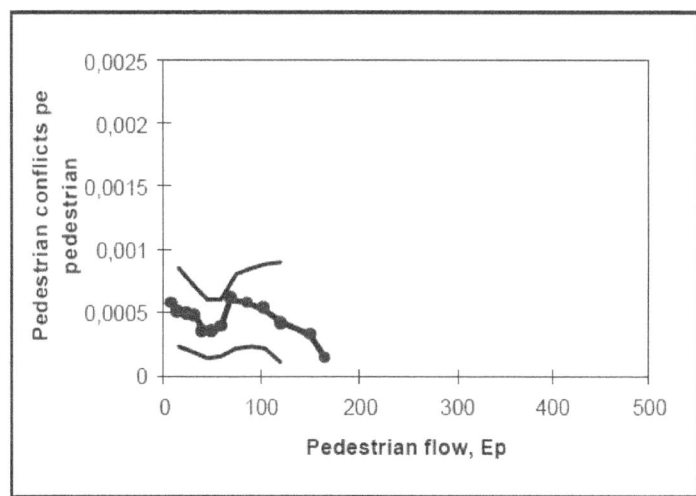

Figure 7. Pedestrian conflicts per pedestrian versus pedestrian flow for approaches without Zebra crossing. (Moving average line (RPF) with an estimated 80% confidence interval.)

Again the approaches with zebra crossings are associated with higher conflict rates. Approaches without zebra crossings have generally much lower pedestrian flow in this data set. This is quite natural, since a high number of pedestrians is a major argument for installing zebra crossings. The shape of the RPF below 100 pedestrians per hour seems identical for approaches with and without zebras even if the irregularities are insignificant for both curves. The difference in the conflict rate is, however, significant between approaches with and without zebra marking.

With and without refuge

A common countermeasure for improving the situation for pedestrians is a refuge. The idea is to both decrease the speed of approaching vehicles and to make the crossing easier for the pedestrian by enabling him/her to cross in two steps. Below is a division of the data set in two groups: 161 approaches have refuge and 206 have no refuge in the data set. The RPF, of car exposure for the two groups are presented in figures 8 and 9, respectively.

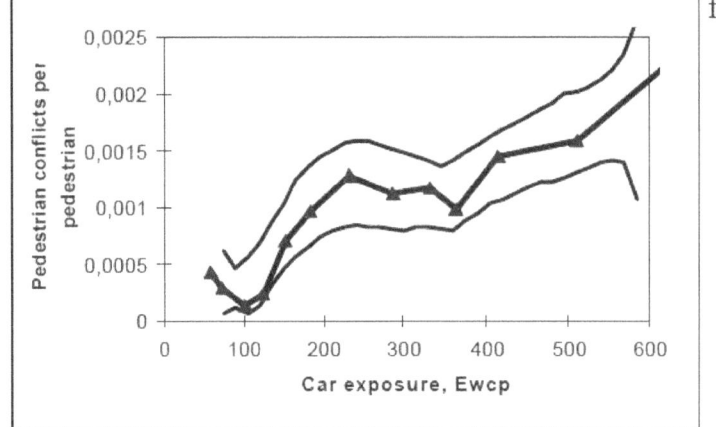

Figure 8. Pedestrian conflicts per pedestrian versus car exposure for approaches with refuge crossing. Moving average line (RPF) with an estimated 80% confidence interval.)

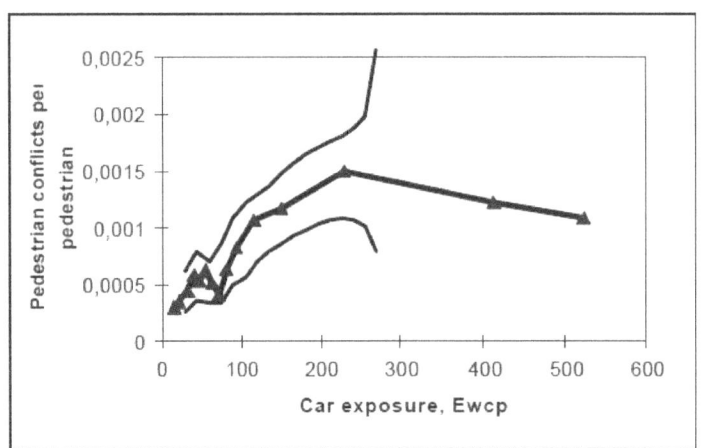

Figure 9. Pedestrian conflicts per pedestrian versus car exposure for approaches without refuge crossing. (Moving average line (RPF) with an estimated 80% confidence interval.)

Despite the fact that most approaches are without refuge, information is gathered over a larger exposure region at locations with refuge. In the low region of car exposure there is just about a significantly lower conflict rate at locations without refuge than at locations with refuge. Remember, however, that locations with refuge normally have zebra crossings as well. The positive safety effect of refuge seems to be stronger than the negative effect of zebra crossing, at least in the lower region of car exposure.

If one distinguishes between with and without refuge, but only for those locations with zebra marking, the results illustrated in figures 10 and 11 are seen.

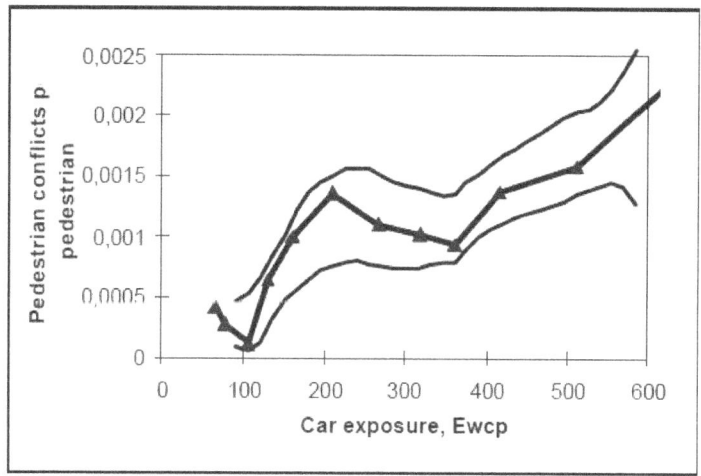

Figure 10. **Pedestrian conflicts per pedestrian versus car exposure for approaches with refuge and zebra crossings. (Moving average line (RPF) with an estimated 80% confidence interval.)**

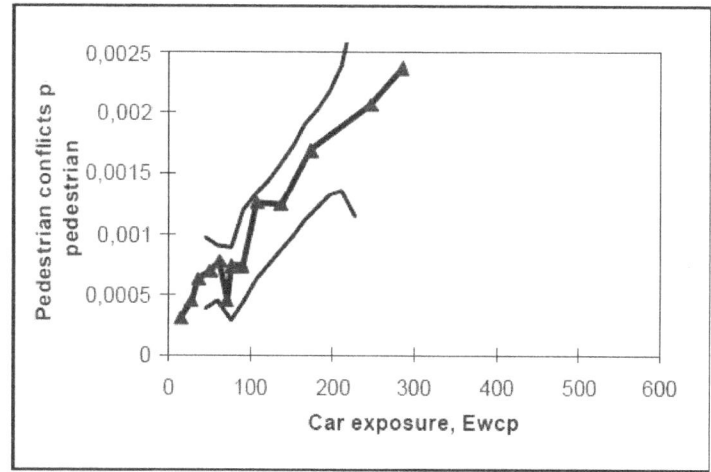

Figure 11. **Pedestrian conflicts per pedestrian versus car exposure for approaches without refuge, but with zebra crossing. (Moving average line (RPF) with an estimated 80% confidence interval.)**

In figures 10 and 11, the positive effect of refuge is slightly increased.

2.2. Small roundabouts

If properly designed, small roundabouts work very well as a speed reducing measure. Experiences of rebuilding a large number of intersections on arterial roads as small roundabouts (diameter between 4 and 25 m (13 and 82 ft)) in England showed that the number of accidents decreased by 30 to 40 percent (NVF, 1984). A large-scale experiment with small roundabouts on arterials in the Swedish town Växjö was carried out in 1991 (Hyden et al., 1992); Hyden, et al, 1995). Twenty non-signalized and one signalized intersection were changed to roundabouts for a period of 6 months. A prerequisite was that the roundabouts only had one lane at each entry and exit and inside the roundabout. This was possible from a capacity point of view, even though the signalized intersection had as much as 22,000 incoming motor vehicles a day. The basic hypothesis was that by using small roundabouts as a speed reducing measure at each intersection speeds would be reduced significantly at the most critical points, i.e., at the intersections, and partly reduced between intersections. The result confirmed the hypothesis; mean speeds through the intersections rebuilt as small roundabouts decreased by 11 to 18 km/h (7 to 11 mi/h) to 30 to 35 km/h (19 to 22 mi/h). Speeds on links between two consecutive roundabouts were reduced between 5 and 10 km/h (3 to 6 mi/h).

Conflict studies indicated that the expected number of injury accidents at the roundabouts would decrease by 53 percent, 66 percent for pedal cyclists and as much as 89 percent for pedestrians. For car occupants no change was predicted. The safety effect was predicted to be the same at the signalized intersection as in the others.

The studies showed that the accident risk was strongly related to the entering speed at these intersections, and that the entering speeds primarily depended on the lateral displacement that drivers had to perform.

Studies also showed that the interaction between car drivers and pedestrians was significantly improved. At the one intersection studied, the number of car drivers that stopped or slowed down to let pedestrians pass increased from 27 percent to 50 percent. The noise level was slightly reduced (by 1.9 to 4.6 dB(A)) at intersections that were provided with roundabouts. Energy consumption and air pollution increased slightly at intersections that were non-signalized in the before situation but reduced much more at the intersection that had been signalized. At non-signalized intersections, there was on average a 5.6 percent increase of CO and a 4 percent increase of NO_x, while at the signalized intersection there was a 29 percent decrease of CO and a 21 percent decrease of NO_x.

The difference in effects between the intersection that was signalized in the before situation and the other was also clearly demonstrated when looking at travel times. On average, car occupants lost 0.12 seconds at non-signalized intersections when they were provided with roundabouts, while there was a time gain of 11 seconds in the signalized intersection. Corresponding figures for pedestrians were no change at non-signalized intersections while they gained 12.5 seconds at the signalized intersection.

Of those 21 intersections that were changed into roundabouts, 5 were kept after the testing period.

Of these, two remained unchanged for 4 years. A long-term follow up was therefore executed at these two intersections. In short, the finding was that speeds had gone up a little but was still much below the original speed. At one of the intersections, the average speed dropped by 50 percent till 4 months after installation of the roundabout and by 34 percent 4 years after, in both cases compared with the before speed. At the other intersection, the corresponding reductions were 51 percent 4 months after and 44 percent 4 years after.

The same kind of finding was true for the accident risk, i.e., there was still quite a significant decrease of risks in the long term, even though it had been slightly reduced compared with the decrease soon after the roundabouts were installed.

Figure 12. Small roundabouts in Växjö.

To conclude, small roundabouts with the aim of not only simplifying driver task in general but also reducing speeds, seem to be very beneficial for pedestrians. At the same time the Swedish study in Växjö clearly indicated that the disbenefits, in terms of increased delay for cars, from changing from give way to roundabout are small and can easily be offset by removing a number of the traffic signals. In Växjö a bit more than 20 intersections were signalized. Most of them could have been replaced by roundabouts with both safety and other benefits as a result. (Today five signals along one of the arterials are changed into roundabouts, all because of the positive assessment made by the city authorities in addition to the research results presented here.)

2.3. Traffic calming

In Sweden, a new publication called *Lugna gatan* (Quiet street) will be launched in 1998. The idea is to provide a handbook covering new traffic safety countermeasures. The aim is also to provide guidance on the network level, thereby enhancing the use of traffic safety countermeasures in a systematic way. This has resulted in a new type of guidelines, where the users are guided more towards a holistic view rather than simple decision rules about what countermeasure to use at a certain location. The general message is in line with the new zero-vision (see section 4.1) namely that the speed at a specific stretch of road should be defined according to the weakest road user.

In urban areas, this means that the speed should be guaranteed not to exceed 30 km/h (19 mi/h) if pedestrians are to be expected. If no pedestrians or bicyclists are expected, the speed could be 50 km/h (31 mi/h) at intersections and 70 km/h (43 mi/h) on the links. The guideline also suggests methods to achieve the low speed at different circumstances.

Environmentally adapted through-roads

Denmark has a great experience with environmentally adapted through-roads. The idea is to create a traffic environment for the weakest inhabitants in a village rather than letting the through-going traffic set the pace. This strategy is a common alternative to bypasses. One important argument for not building bypasses is to hinder a drift of the village center or to drain the village by the through going customers. The evaluation of 21 of these environmentally adapted through roads shows positive effects both on aesthetics and on traffic safety (Vejderektoratet, 1996). The decrease in speed of the traffic has been significant. The decrease in the average speed was 12 km/h (7 mi/h) when round-abouts where used and slightly less where only more "visual" means where used. The increase in travel time was about 10 to 20 seconds since the length of the villages is about 1 km (.6 mi). Since pedestrians crossing the main village street were considered the primary problem, the pedestrians will gain much from the decrease in vehicle speed. If the countermeasures are designed without any physical and intervening ingredient, the effects are not to be expected to be beneficial to pedestrians to any larger degree.

2.4 Project WALCYNG — How to encourage WALking and CYcliNG instead of shorter car trips and to make these modes safer

Introduction

WALCYNG was a project funded by the European Commission under the Transport RTD Programme of the 4th Framework Programme. It had partners from eight countries and was coordinated by the Department of Traffic Planning and Enginneering in Lund, Sweden.

The increase in car traffic has become a threat to the quality of life in urban areas. Accidents and other safety related problems are examples, and local emission problems are another. More generally, it is evident that car traffic in most urban areas has grown so much that many important aspects of urban life are inhibited to such an extent that the question of sustainability has become an important topic.

Promoting walking and cycling as an alternative to short car trips is seen as one important way of decelerating the increase in car use in more densely inhabited areas. Pedestrians and cyclists produce no major threat to other road users, nor do they pollute the environment with fumes and noise. Besides pedestrians and cyclists provide the best opportunities of enhancing qualities of life in general in urban areas.

The purpose of WALCYNG was to sort out conditions and measures which may contribute to replacing short car trips with walcyng (walking and cycling). WALCYNG applies a Marketing Model, in the project formalized in four main parts:

1. Information policy: One has to collect information about potential and practising customers so that the preconditions for the behavior they should choose could be made attractive.

2. Product and distribution policy: Adequate and attractive technical solutions are worked out and considered thoroughly so that they will meet customers' and potential customers' needs.

3. Incentive and pricing policy: One also must provide incentives given by the society, institutions, companies, etc., on all levels, both to encourage walking and cycling and to discourage the use of cars for short trips.

4. Communication policy: Users and potential users must be informed that their needs and interests are taken into consideration, on the product and distribution side, as well as on the incentive side. The product must be displayed and be given a positive image.

Main findings

a. What is known about the target groups and their situation?

Many car trips are quite short; a change from car to walking or cycling for trips shorter than 3 to 5 km (2 to 3 mi/h), could replace half of all car trips in many European cities. Trip chains could only explain some of the car use on short trips. Important differences are found between men and women, young and old, car owners and people without a car, workers and non-workers.

A lot of products and efforts were identified. They were divided into four different types:

(1) Personal products, i.e., products that are appropriate to wear or to be used for help or comfort, for weather protection, for carrying things, for security or items to facilitate walcyng.

(2) Vehicle products belong primarily to the bicycle or could be attached to it (bags, lamps, computers, etc.).

(3) The Road and infrastructure category deals with design and maintenance on net level of links, crossings, parking facilities, and intermodal points, as well as restrictions for motorized traffic.

(4) The aim of societal efforts (e.g., media, politicians, officials, and companies) is to reach certain attitudes and behavior among the public supporting walcyng interests and/or discouraging the use of cars. The means can either be persuasive or forcing.

The experienced problems of walcers were analyzed along the dimensions of social climate (e.g., the low status of walcyng), health (e.g., cycling is good for health but cannot be done without a baseline health condition), comfort (e.g., important with special provision of benches, waste-baskets, shelters, and public toilets), subjective safety (e.g., pedestrians should be separated from both cars and cyclists, walcers should not be too much isolated, especially if the illumination is poor), mobility (e.g., the bicycle network must be continuous and of good quality), aesthetics (e.g., pedestrians and cyclists have time to look around and really get to know the environment), and financial advantage.

Even though it is a fact that the more cyclists there are in a country the lower the accident rate, increased walking and cycling would result in a considerable increase in accidents if no strong action is taken. WALCYNG presents eight recommendations valid in most European countries. In most cases they can be implemented with reasonable costs in a short term. The far most important measure WALCYNG wants to highlight is a strategy to achieve a maximum speed of 30 km/h (19 mi/h) on streets where walcers are present.

b. What is known about the preconditions for WALCYNG?

There are a lot of benefits associated with walking and cycling: Health aspects are important benefits of walking as well as of cycling. For walking, environmental aspects and getting fresh air are additional important benefits. Surprisingly, in interviews, environmental aspects are not mentioned very often as positive aspects. Cycling is fun, gives you good exercise, and is very convenient. Even though there are many benefits involved in walking and cycling, walcers meet a lot of barriers and obstacles, e.g., lack of facilities to transport heavy things, hilly topography, bad weather, polluted air, as well as infrastructure barriers such as insufficient road cycle network, unsafe crossings, parked cars on the pavements, high curb stones, etc.

The figure below gives an example from an attitude survey carried out in four countries. It presents the main barriers for walking.

A Norwegian stated preference-study indicated that the trips to work and to sports and exercise are easiest replaceable by bicycle, while grocery shopping trips could easiest be replaced by walking.

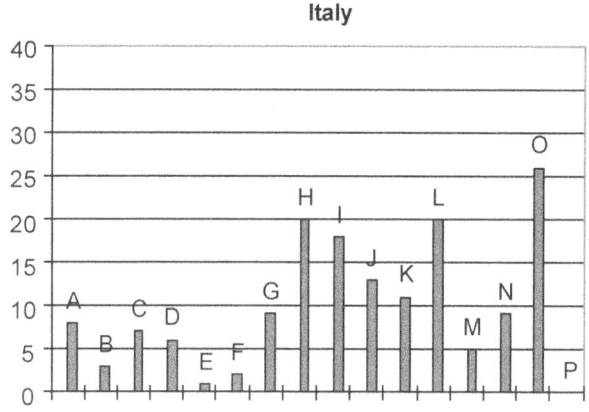
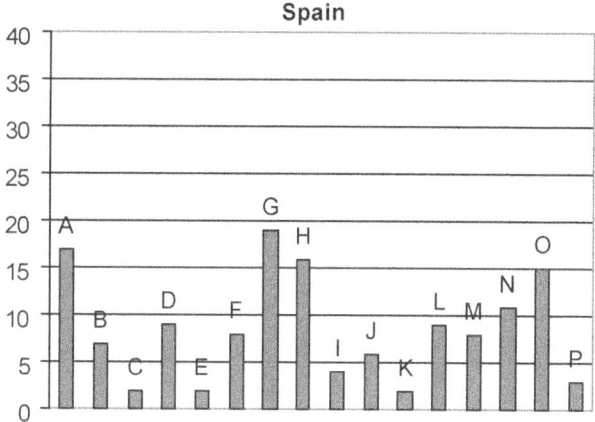
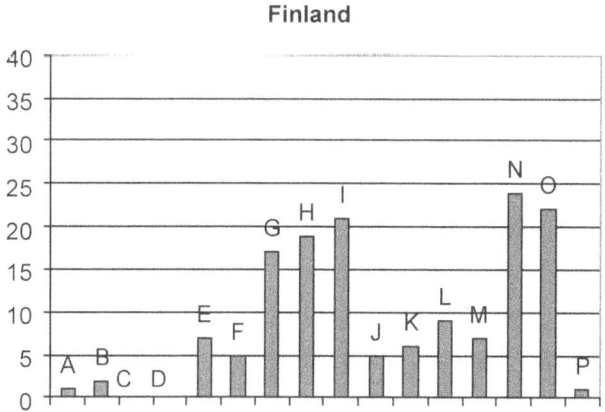
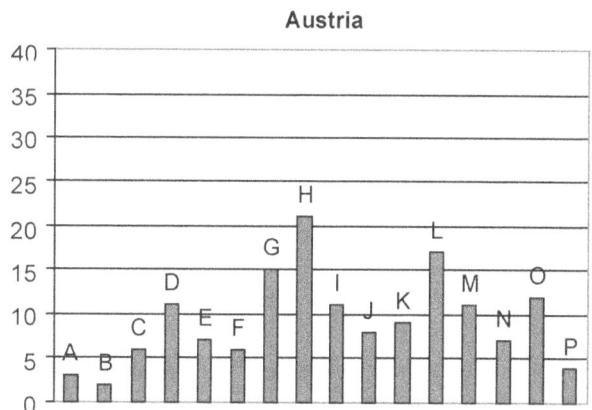

A	=	bad upkeep of pedestrian ways	I	=	Weather
B	=	lack of pedestrian ways	J	=	high speed of traffic
C	=	Subways	K	=	unattractive surroundings
D	=	long detours	L	=	car noise and pollution
E	=	cyclists on pavement	M	=	Ruthlessness of car drivers
F	=	badly designed traffic lights	N	=	Laziness
G	=	feeling of insecurity at night time because of badly lit paths	O	=	walking takes a lot of time
H	=	nonability of transporting heavy things	P	=	other

Figure 13. Barriers for walking in Finland, Austria, Italy, and Spain. The attitude surveys.

In the same attitude survey, the respondents were asked to identify measures that could reduce barriers and improve walking and cycling conditions. Most respondents see infrastructural and political measures as most important. Among the infrastructural measures are:

- More footpaths and cycling lanes.
- Wider sidewalks.
- Subways and crossings.
- Smoother road surface.

The most important political measures are:

- Prohibit cycling on sidewalks.
- City centers free of cars.
- Priority for walcers at crossings.
- Different measures to reduce car traffic, like increased gasoline prices, fees for parking spaces, restrictions on car driving, etc.

(1) What incentives and disincentives should be provided for car drivers to make them walk and cycle instead of using the car for short distances?

Incentives are important instruments to influence travel behavior in a walcyng-friendly way. An important conclusion of WALCYNG was that incentive strategies should play a more important role in the future, both providing a walcyng-friendly infrastructure and atmosphere, and giving economic incentives to make walcyng more competitive compared with using the car.

(2) How good is communication with the target groups and how can it be improved

Communicative measures (e.g., information campaigns, and advertising) should consider a segmentation of the market and different characteristics of different target groups. The sender of the message may vary, as long as he or she is credible. The most important characteristics of good communication are summarized in the final report.

(3) How should researchers and practitioners prepare themselves for the structural difficulties they will meet with a topic that so far is considered of inferior importance?

One important aspect of promoting walcyng is the problems one can expect to meet in terms of partly naive arguments against walcyng from various representatives of governmental institutions, private institutions, or certain individuals working in such institutions. It is the idea of WALCYNG that mentally dealing with the expected problems helps one to react in a more relevant and objective way when they arise. The involved person will thereby be better prepared and resistant. The final report presents relevant counterarguments for all types of arguments.

(4) How to raise and sustain the importance of walking and cycling as transport modes?

Lobbying plays an important role in promoting walcyng. The most important aspects in connection with lobbying are, e.g., to become part of the political sphere, to achieve the assistance of media to influence people's attitudes and behavior, and to consider the fact that the effectiveness of one's efforts is linked to power, like economical, representative, psychological, power through access to resources, etc. It was concluded that successful lobbying needs a network of cooperating partners.

The WALCYNG Quality Scheme

One of the main goals of WALCYNG was to produce an evaluation scheme; the Walcyng Quality Scheme (WQS). It should allow an assessment of different policy activities in the area. The WQS is designed as an interactive software that can be used for obtaining and evaluating information about the preconditions for walcyng in a certain area of interest (a target group, a type of product, a route, a neighbourhood, a city, a country, etc.). The WQS should on the one hand remind the compilers of all relevant aspects to be considered both when assessing given preconditions for walcyng and when developing measures. In this respect, the WQS resembles a checklist. On the other hand, the WQS has a comparative and an analytic character because the quality of the aspects that should be considered has to be assessed as well.

2.5 Car - Pedestrian interaction at zebra crossings

Várhelyi (1996) has carried out an indepth study of driver and pedestrian behavior at a non-signalized zebra crossing, with special emphasis on drivers' choice of speeds when approaching the crossing.

The study was made on a 7-m (23-ft) wide arterial road in the city of Lund, Sweden. Sight conditions were quite good. Speeds were recorded with a radar gun, and the events were video-recorded at the same time. The study focussed on so called interactive situations, defined as situations where a vehicle was approaching the crossing — within a distance of 70 m (230 ft) — when there was a pedestrian present at the same time.

The results showed that of 824 interactive situations the pedestrian crossed before the vehicle in only 5 percent of the cases. This situation only changed when the pedestrian was more than 4 seconds ahead of the vehicle. Then vehicle speeds were lower because the drivers thought that "it was no use in trying to pass before the pedestrian." This also made it more probable that the pedestrian actually passed before the vehicle.

Vàrhelyi also defined "critical interactions." Those were the interactions where the vehicle and the pedestrian in theory could reach the meeting point exactly at the same moment. The table below shows the speed behavior of drivers in these critical interactions.

Table 3. Speed behavior in critical interactions at a zebra crossing.

Speed behavior	Number of drivers	Percent (%)
Reduce speed	16	11
Braking	24	16
Keep the same speed	86	57
Accelerating	25	16
Total	151	100

The table shows that drivers keep the same speed or accelerate in 73 percent of the cases, while only 27 percent of the drivers are slowing. The results indicates that there seems to be a driver strategy to maintain speed — or even accelerate — even with speeds above the prevailing speed limit of 50 km/h (31 mi/h) to communicate to the pedestrian that he/she has no intention to stop and let the pedestrian pass. This communication in the form of acceleration took place 40 to 50 m (131 to 164 ft) away from the crossing. From the driver's point of view, the strategy works quite well; the vehicle passes before the pedestrian in almost all cases. From a pedestrian point of view the strategy is less attractive. Besides, the present law clearly states that a driver who is approaching a zebra crossing where a pedestrian is present has to slow down so not to put the pedestrian in danger.

This kind of driver behavior was also demonstrated through mean speed profiles produced for vehicles approaching the crossing. At noncritical interactions, the mean speed was significantly lower than both critical interactions and situations where no pedestrians were present.

To overcome the problems at zebra crossings, there are intentions in Sweden to change traffic rules so that there will be an absolute, and clearly stated, requirement to yield for pedestrians at zebras.

3. The use of new pedestrian facilities

3.1. Detection of pedestrians at signal-controlled intersections

In all the Scandinavian countries, special pedestrian facilities are installed in most signal-controlled intersections. The pedestrian phase is, unlike the case in United Kingdom, often linked to the vehicle phase. This means that a green pedestrian signal does not guarantee the absence of conflicting vehicles. Elvik et al (1995) found that the installation of traffic signals that allow for conflicting vehicles in the green phase for pedestrians produce an 8-percent increase in pedestrian accidents while the installation of signals that do not allow for conflicting vehicles produce a 30 percent decrease in pedestrian accidents.

Quite often, the pedestrian phase is triggered by the vehicle detector or a constant demand is programmed. In Finland and Sweden, it is common to have pedestrian push buttons at most intersections. Push buttons are generally considered pedestrian-unfriendly. In Finland, there are several installations in Helsinki where radar detectors are used to detect pedestrians approaching a pedestrian crossing (Kronborg and Ekman, 1995). Radar detection is especially used when there is a high probability that a pedestrian approaching the crossing also will cross the street. A few tests have also been done in Sweden and Norway to extend the green by using radar detectors. One problem noticed is that cars standing close to the pedestrian crossing can be detected as pedestrians. The extension of the green time does not solve any real safety problem. Detecting pedestrians to prevent pedestrians from walking against red signal is more important from a safety point of view (Kronborg and Ekman, 1995; Almqvist, 1996).

In a joint European study (Ekman and Draskozy, 1992), trials with microwave detectors to trigger the traffic signal were carried out. In one trial in a small town (Växjö), detectors where mounted to detect all approaching pedestrians. The detectors were connected to the pushbuttons, thereby having the same effect as pushing the button.

The result showed a significant reduction of red walking among the pedestrians, simply because it was more often green when the pedestrians arrived at the pedestrian crossing. The effect was especially strong for pedestrians crossing the minor road. The effects on vehicle traffic were negligible since the traffic signal program was very vehicle-friendly to start with. The false detection of pedestrians approaching the crossing without any attempt to cross was not found to be any major problem. It was not bigger than the present problem with pedestrians pushing the button without waiting for the green light either by walking against red or by crossing the other street in stead.

In general, if the traffic engineer used the pedestrian detection information more comprehensively, the results could have been even more favorable both for pedestrians and car occupants. From a vehicle perspective, no green time need to be spent on pedestrians if no pedestrians are present; and from a pedestrian point of view, delay was reduced.

The conclusion of this study is:

- It is possible to detect approaching pedestrians in a reliable way.
- Significant reductions in red light violations can be achieved.
- False detection was no major problem.

In this intersection, no alternative signal strategies were tried. A more sophisticated use of detection data could have led to further improvements. Even if proper counting is hard to do with a standard microwave detector, it could be used to produce rough estimates of the number of approaching pedestrians. This was shown at a test site in Oporto in Portugal (VRU-TOO,1995).

3.2. Relevant warning system

At one intersection in Växjö, the local authorities encountered a problem with low respect for an ordinary zebra crossing. The problem was that many children passed the intersection to and from school and that cars tend not to stop and yield for the children. The present warning sign had no effect. A new big warning sign, activated by the presence of pedestrians was installed.

Figure 14. Warning sign in the city of Växjö (STANNA FÖR GÅENDE = Stop for Pedestrian) (Towliat, 1997).

The results indicated a remarkable increase in the number of vehicles stopping to let pedestrians cross the street. Before the new sign was introduced, about 12 percent of arriving cars stopped when pedestrians where present. Right after the sign was introduced, 50 percent of the cars stopped. After 1 year, more than 50 percent of the cars stopped (Towliat, 1997; Towliat working material, 1998). The positive results are explained by both the fact that vehicle speed was reduced, before the installation of the new sign, to below 50 km/h (31 mi/h) and that the new sign gave a relevant and new type of warning to the drivers. The sign was not activated very often since pedestrian volume is low. This means that even if a car driver passes this intersection once a day it might take weeks until he/she sees it activated. This could explain that the sign could still have an effect still after a year.

3.3. Warning lights mounted at the roadways

Another example of a relevant warning system is seen in a small experiment in the City of Helsingborg where small warning lights where mounted at the roadway in two signal- controlled intersections. The lamps used where of the same type often used on airport runways. At one intersection, the lamps were intended to make the turning vehicles pay attention to on-coming, straight going, vehicles in the same signal phase. In the other intersection, the lamps were used to remind turning vehicles that crossing pedestrians have the right of way.

The results from the evaluation (Ekman, 1995) was:

Figure 15. Warning lights mounted at the roadways in the city of Helsingborg, Sweden.

- The lamps worked technically well.

- At one of the intersections, a significant safety effect was found.

- At the other intersection, the safety problem from the beginning was so small that no major improvements in safety were possible.

- The system could be further improved if pedestrians could be detected.

3.4. Painted pre-marking at zebra crossings in Stockholm

In Stockholm a serious problem was addressed at zebra crossings with traffic in more than one lane in each direction. It is considered a severe problem that overtaking often takes place just before these zebra crossings. If one driver stops to let a pedestrian cross, it often results in a dangerous situation if the driver in the other lane does not stop. The pedestrian and the overtaking driver have very little chance to see each other since the stopped car hinders the view. The pre-marking was intended to create better visual circumstances in this situation by encouraging drivers to stop a bit further ahead of the zebra crossing

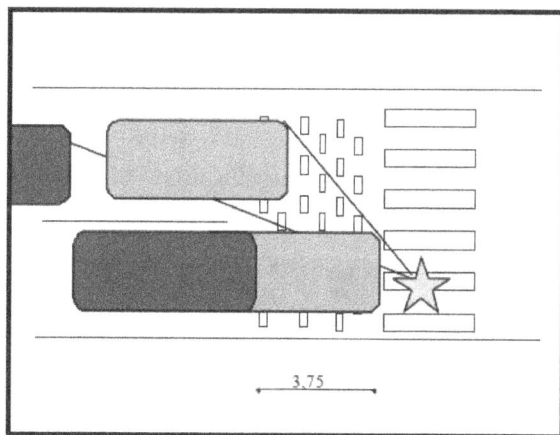

Figure 16. Sketch of Pre-marking illustrating the intended increase in sight distance.

If cars stop before the pre-marking, the sight distance would increase significantly.

The pre-marking was evaluated with conflict studies, speed measurements, and behavior studies before and after the installation of the marking. (Towliat and Ekman, 1997)

The general finding was that the pre-marking did not significantly improve the safety situation for pedestrians crossing the street at the zebra crossing. Only 6 percent of drivers stopped before the pre-marking (4 m (13 ft) before the zebra marking). In the before situation, 4 percent stopped 4 m (13 ft) before the zebra marking, i.e., where the pre-marking later was located.

The conflict result showed no change in the number of serious conflicts. The conflict studies confirm, however, the initial picture of the problem for pedestrians crossing wide streets with several lanes and heavy car flow: An overtaking car was involved in about 50 percent of the serious conflicts. The speed measurement on free cars showed a minor increase. The increase was small and may well have been caused by something other than the countermeasure.

The reason that the pre-marking did not lead to any major improvement could be either that the countermeasure was too weak (only advisory) or that it was compensated by an increased risk taking by the pedestrians.

3.5 Fluorescent caps on first-class pupils

Since the young children, 6 to 7 years of age, are very inexperienced in traffic, new school pupils are very vulnerable. Fluorescent caps were given to many school children in Sweden, when arriving at school to enhance their visibility when walking to and from school. A similar project has been carried out in Denmark, where the small children where given colourful rucksacks or caps. In Sweden, Dahlstedt (1995) carried out an evaluation of the effect of fluorescent caps on first-class pupils.

The main purpose of the evaluation was to test a hypothesis about the so-called "Superman effect." There was a strong criticism from several researchers that the fluorescent caps might result in a false feeling of safety from the pupils. It was believed that the pupils might think, "if I use this "magic cap," no car could hurt me."

Figure 17. Pupils with flourescent caps.

The "superman effect" was studied by means of interviews in 46 classes, in which the children had been given caps the previous term. The results showed some examples of what could be expressions of an imagined invulnerability, but closer analyses indicate that these examples, rather was a result of bad introduction of the fluorescent caps, than logical conclusions, from the children them selves.

The study of the drivers' behavior was rather limited but it indicates that at least at some locations the caps contributed to a somewhat lowered approach to the crossing. The caps also seems to make the drivers observe the children significantly better, but it could not be shown that this affects the readiness of drivers to give way to the child pedestrians.

Our conclusion of the use of fluorescent caps is that they may be of good use if not introduced for traffic safety reasons. Then one could avoid that a false feeling of safety for both children and parents is introduced.

3.6   Ultraviolet light

Some years ago in the work within DRIVE and Prometheus, big hope was put on the use of ultraviolet light to help the car driver to see pedestrians better during night. The benefit with the UV-light is that one does not need to dim the headlights when meeting other vehicles. The UV-light gives good reflection on light clothes or special fluorescent painted material. A major problem is of course that not many pedestrians wear bright colors.

The development of the UV-light has more or less stopped for two different reasons. The first is that it seems very difficult to produce lamps at a reasonable price. The other reason is that some researchers claim that the use of UV-lights when driving in the dark will lead to higher speeds and that elderly drivers will start driving during the night to a larger degree. Thereby the number of accidents will increase if the use of UV-light became common. This compensating effect has been found on ordinary reflector poles that are used on smaller rural roads in Sweden.

4. Ongoing and future research in Scandinavia

4.1. Implementation of "Vision Zero"

In Sweden the National Road Administration has formulated a new and elaborated target for the future traffic safety work. It is called the "Vision Zero." The vision is that in the future no severe or fatal accidents shall occur in traffic. Even if this is unrealistic in one way, the new vision has meant a lot for the attitude towards traffic safety issues. It has put a very strong emphasis on the responsibilities of the accidents on our roads.

On October 9, 1997, the Road Traffic Safety Bill founded on "Vision Zero" was passed by a large majority in the Swedish Parliament. This represents an entirely new way of thinking with respect to road traffic safety.

Vision Zero is conceived from the ethical base that it can never be acceptable that people are killed or seriously injured when moving within the road transport system. It centers around an explicit goal and develops into a highly pragmatic and scientifically-based strategy which challenges the traditional approach to road safety.

The long-term goal is that no one will be killed or seriously injured within the Swedish road transport system.

A new approach to road safety

For many years, the emphasis in traffic safety work has been to encourage the road user to respond, in an appropriate way, typically through licensing, testing, education, training, and publicity to the many demands of a man-made and, increasingly complex traffic system. Traditionally, the main responsibility for safety has been placed on the user to achieve this rather than the designers of the system.

The Vision Zero approach involves an entirely new way of looking at road safety and the design and functioning of the road transport system. It involves altering the emphasis away from enhancing the ability of the individual road user to negotiate the system to concentrating on how the whole system can operate safely. Also, Vision Zero means moving the emphasis away from trying to reduce the number of accidents to eliminating the risk of chronic health impairment caused by road accident.

The traffic system has to adapt to take better account of the needs, mistakes, and vulnerabilities of road users. The level of violence that the human body can tolerate without being killed or seriously injured forms the basic parameter in the design of the road transport system. Vehicle speed is the most important regulating factor for safe road traffic. It should be determined by the technical standard of both roads and vehicle so as not to exceed the level of violence that the human body can tolerate.

Vision Zero accepts that preventing all accidents is unrealistic. The aim is to manage them so they do not cause serious health impairments. The long-term objective is to achieve a road transport system that allows for human error but without it leading to serious injury.

While the concept envisages responsibility for safety among the designers and users of the system, the designer has the final responsibility for fail-safe measures.

System designer has primary responsibility

System designers are responsible for the design, operation, and the use of the road transport system and are thereby responsible for the level of safety within the entire system. Road users are responsible for following the rules for using the road transport system set by the system designers. If the users fail to comply with these rules because of a lack of knowledge, acceptance, or ability, the system designers are required to take the necessary steps to counteract people being killed or injured.

Vision Zero sets out the operational principles that would need to be taken up by citizens, decision makers, public authorities, the market and mass media if the strategy is to be effective.

VISION ZERO: OPERATIONAL PRINCIPLES:

- At a political level, not allowing road traffic to produce more health risks than other means of transportation or other major technological systems.

- At a professional level, seeing serious health loss caused by traffic accidents as an unacceptable quality problem of products and services connected with road transportation.

- At an individual level, viewing serious health loss as unacceptable, being aware of what it takes to create a safe system, and playing an active part in placing demands on society and manufacturers for safe road traffic.

Action in a variety of fields is needed to produce a safe road system

VISION ZERO: ACTION STRATEGY:

- To prevent accidents leading to serious injury.
- To reduce the severity of injury in the event of an accident.
- To ensure that the severity of injuries received is minimized through efficient rescue service, health care, and rehabilitation.

A result-based action program for safe road traffic within the principles outlined above will be defined by the Swedish agencies for future road safety work which should lead to the realization of Vision Zero in the long run.

In the next 10 years, it is estimated that it should be possible to reduce the number of fatalities by a quarter to one third.

4.2. Speed Limiters for controlling vehicle speeds

The main conclusion of the Zero Vision regarding pedestrian safety is that low motor vehicle speeds at every encounter between a motor vehicle and a pedestrian must be assured. The maximum speed is set to 30 km/h (19 mi/h). All available empirical evidence supports the argument strongly: an impact speed of 55 km/h (31 mi/h) in a pedestrian accident is fatal with almost 100 percent probability; while at an impact speed of 30 km/h (19 mi/h), the risk of being killed is reduced to less than 10 percent (Pasanen, 1992). A reduction of the average travelling speed in a flow by as little as 10 percent on average reduces the fatality risk by as much as 35 percent (Carlsson, 1980). A travelling speed of 30 km/h (19 mi/h) will in most cases result in an impact speed, which is considerably lower than 30 km/h (19 mi/h), thus producing very small risk of serious accidents for pedestrians.

Lots of efforts have been made to redesign the road infrastructure to reduce speeds at intersecting points as is shown in other parts of this report. The main problem with this approach, limiting the scope considerably, is that the cost of implementation is very high if measures are introduced on a large scale, which is a necessary prerequisite if large-scale effects is the objective.

Gradually, research on the effects of speed control in private cars is growing, particularly in Sweden. Some different concepts are tested, e.g., a system producing a beeping signal when the speed limit is exceeded. An experiment was carried out in the Swedish city of Umeå where 100 cars were equipped with a system that was activated at two road sections with a speed limit of 30 km/h (19 mi/h) outside schools.

At the University of Lund in Sweden, research has been carried out for 10 years on the so called Speed Limiter (SL) concept. The vehicle can simply not exceed the existing speed limits. When the speed is reaching the speed limit the acceleration stops, and the vehicle keeps the speed of the speed limit as long as the driver presses the accelerator. At lower speeds, the vehicle is working as any vehicle without a Speed Limiter.

In an experiment in the city of Eslöv, 30,000 inhabitants, a system was tested where information about existing speed limit (50 km/h (31 mi/h)) was transferred to the vehicle via transponders located at each speed limit sign at the city borders (Vägverket, 1997). The maximum speed of the vehicle was then automatically set to 50 km/h (31 mi/h). Twenty-five drivers were included in the test. They were all driving for at least 2 months. Studies were carried out and included driving before the SL and at the end of the period (with SL), manual observations from the car, automatic measurements via a data-log in the vehicle, and interviews.

Interview results

The interviews were focused on drivers acceptance of the SL-function in their daily life. The distribution of answers are presented below:

Question: *How often do you appreciate the existence of the SL-function when using your car?*

Every time	Almost every time	Often	Seldom	Never
70%	24%	6%	0%	0%

Question: *How was your feeling with the feedback from the accelerator pedal?*

Excellent	Good	All right	Poor	Disgusting
12%	53%	35%	0%	0%

Question: *How did you like the SL-function you have been using, compared with your expectations?*

More positive	As expected	More negative
70%	18%	12%

Question: *What is your attitude to introducing the SL-function at all different speed limits in urban areas (30, 50, and 70 km/h (19, 31, and 43 mi/h))?*

Very positive	++	+	0	-	--	Very negative
26%	21%	33%	13%	7%	0%	0%

Question: *What is your experience of the SL-function as...*

	Agree	+	0	-	Disagree
..a sense of security?	12%	35%	35%	12%	6%
..a support?	35%	53%	6%	6%	0%
..an unpleasant control?	0%	6%	6%	41%	47%
..a source of irritation?	0%	12%	12%	35%	41%

Question: *Do you think that beeping signals and/or flashing lamps inside the vehicle would have the same effect as the SL-function to keep the speed limits?*

No	?	Yes
87%	13%	0%

Question: *Are you willing to pay out of your own pocket to install an SL-function in your own car?*

No	Yes, if it is integrated	Yes	SEK
29%	18%	53%	1,000 (approx. 125 USD)

Some biases of the results because of psychological effects must be expected. A small group of testers of a new system that is raising a lot of interest from the media was addressed. It is therefore likely that the testers are more in favor of the system than what can be expected later on. Still the results are convincing with tendencies that are very strong.

Driving behavior

Inappropriate interaction with vulnerable road users (average number of errors made by the drivers).

Before (without SL)	After (with SL)	Difference
2.6	0.5	-2.1

Interactions with other road users together with several other behavioral parameters observed indicates a better interplay and calmer way of driving like smoother accelerations.

Driving patterns

Time consumption recorded for the test route.

Before (without SL)	After (with SL)	Difference
1426 sec	1487 sec	+61 sec (4.3%)

Speed profiles

The studies indicate lower and more uniform speeds when the SL is used. The speed reducing effect of the SL in mixed traffic is biggest where the possibilities of exceeding the legal speed limit are most pronounced.

To conclude, speed control in vehicles is slowly being accepted as a possible way of solving problems with excessive speeds. A great deal of research is, however, still warranted on feasibility questions, type of concept (e.g., if the system should be based on warning or on voluntary or involuntary use of Speed Limiters), effects on safety, and other aspects like efficiency, noise, and air pollution. In a 4-year period, starting in 1998, a large-scale experiment on dynamic speed adaptation is going to take place in Sweden, where all these questions will be studied empirically.

In theory, speed controlling systems have very significant safety effects. A system that prevented all speeding would reduce injuries by 20 percent and a maximum speed of 30 km/h (19 mi/h) where vulnerable road users interact with cars — as proposed by the Swedish National Road Administration as a part of the Zero Vision strategy — would reduce fatal and severe injuries for pedestrians and cyclists by at least 80 percent.

Since the major problem for pedestrians in urban areas is the high speed of cars, the speed limiter offers one of the most powerful tools for improving the situation for pedestrians and bicyclists in urban areas.

5. Conclusion

Even though the pedestrian situation in Scandinavia is relatively good compared to other countries and bicycle planning in Scandinavia is considered as an important part of today's traffic planning, these road user groups have a hard time in the modern traffic environment. Even if we use the public transport or private car, we act as pedestrians in most trips, at least part of the trip. Pedestrians are by nature vulnerable and pedestrians do not consider them selves as road users. The basic problem is that traffic planning still is dealt with as a vehicular issue only in many cases.

The conclusion from most recent research is that there is a need to guarantee either complete separation between pedestrians and vehicular traffic, or create good conditions for proper interaction between the pedestrian and the driver. Good condition for interaction could be achieved if the vehicle speed is below 30 km/h (19 mi/h) and the layout gives a simple and clear arena for communication.

The insight into the basic problems for pedestrians could either be used to further blame and restrict the pedestrians or it could be used to understand the conditions under which traffic planning has to be done. Many of these problems are rather difficult to eliminate, good solutions are normally a result of countermeasures that addresses several of these basic problems.

6. References

Almqvist, S., Kronborg P., (1996), Trafik och trafiksignaler med inriktning på trfikantbeteende. Trafiksäkerhetsstudie i tre signalreglerade korsningar. (*Road User Behavior at Three Traffic Signals*), Department of Traffic Planning and Engineering, Lund Institute of Technology, Sweden.

Berntman, M., Modén B., (1996), Malmötrafikens problem i ett sjukhus perspektiv - En medicinsk uppföljning av trafikskadade två år efter olyckan. (*The Traffic Safety Problem in Malmö from a Medical Perspective — A Follow-up Study 2 Years After the Accident*). Malmö Gatukontor. Sweden.

Carlsson, G., (1980), Beräkning av förväntad olycksreduktion vid förbättrad efterlevnad av gällande hastighetsgränser. (*Estimation of Expected Effects at a Higher Degree of Compliance with Speed Limits*). VTI Meddelande 222, Linköping; Sweden.

Dahlstedt, S, (1995), Fluorescerande kepsar på försteklassare - några utvärderingsresultat. (*Fluorescent Caps — Some Results*) VTI rapport Nr 405 VTI, Sweden.

Ekman, L., (1988), Bulletin 76. Fotgängares risker på markerat övergångsställe jämfört med andra korsningspunkter, (*Pedestrian Risk at Zebra Crossings Compared to Other Crossing Points*). Department of Traffic Planning and Engineering, Lund Institute of Technology, Sweden.

Ekman, L., (1996), *On the Treatment of Flow in Traffic Safety Analysis*, Department of Traffic Planning and Engineering, Lund Institute of Technology, Sweden.

Ekman, L., and Draskóczy M.,(1992), *Trials with Microwave Detection of Vulnerable Road Users and Preliminary Empirical Model Test*. ITS Working Paper 336, Leeds, UK.

Kronborg, P., Ekman L., (1995), *Traffic Safety for Pedestrians and Cyclists at Signal-Controlled Intersections*. TFK. Stockholm, Sweden.

Elvik, R. m fl. Utkast til reviderte tiltakskapitler i Trafikksikkerhetshåndboka: Trafikkregulering. Transportøkonomisk Institutt, arbeidsdokument, (*Working Document for the Traffic Safety Handbook*) TST/0687/95. Oslo; 1995.

Hydén, C., (1987), Bulletin 70, *The Development of a Method for Traffic Safety Evaluation: The Swedish Traffic Conflicts Technique*. Department of Traffic Planning and Engineering, Lund Institute of Technology. Sweden.

Hydén, C., Odelid K., Várhelyi A., (1995), Effekter av generell hastighetsdämpning i tätort, Huvudrapport. (*The Effects of a General Speed Reduction in Urban Areas, Main report*) Department of Traffic Planning and Engineering, Lund Institute of Technology. Sweden.

Hydén, C., Odelid, K., Várhelyi, A., (1992), Effekten av generell hastighetsdämpning i tätort. Resultat av ett storskaligt försök i Växjö. (*The Effects of a General Speed Reduction in Urban Areas, Result of a Large-scale Experiment*), Department of Traffic Planning and Engineering, Lund Institute of Technology. Sweden.

Jacobs, G. D. & Wilsson, D.G., (1967), *A Study of Pedestrian Risk in Crossing Busy Roads in Four Towns*, Road Research Laboratory, Crowthorne, UK.

LTH/VTI (1996), Fotgängares och cyklisters singelolyckor (*Single Accidents Among Pedestrians and Cyclists*). Bulletin 140, Department of Traffic Planning and Engineering, Lund Institute of Technology. Sweden.

NVF, 1984.: Rundkjoringer, (*Roundabouts*) - rapport nr 27.

Pasanen, E., (1992), *Driving Speeds and Pedestrian Safety; A Mathematical Model*. Publication 77. Helsinki University of Technology. Finland.

SCB, (1997), Trafikskador 96. (*Police Reported Accidents in Sweden 1996*), Sweden.

Towliat, M., and Ekman L., (1997), Föregångsmarkering - en utvärdering av trafiksäkerhetseffekten av föregångsmarkering vid markerat övergångsställe på breda gator. (*A Traffic Safety Evaluation of "Pre-marking" at Pedestrian Crossings at Wide Streets in Stockholm*). Department of Traffic Planning and Engineering, Lund Institute of Technology. Sweden.

Towliat, M, (1997), Trafiksäkerhetsproblem och åtgärder för gång- och ykeltrafikanter o mötespunkter med bilister. (*Traffic Safety Problem for Pedestrians and Cyclists at Crossing Points*), Department of Traffic Planning and Engineering, Lund Institute of Technology. Sweden.

Towliat, M, (1998), *Working Material on the Effect on Relevant Warning System*. Department of Traffic Planning and Engineering, Lund Institute of Technology. Sweden.

Várhelyi, A., (1996), Bulletin 142, *Dynamic Speed Adaptation Based on Information Technology — a Theoretical Background*. Department of Traffic Planning and Engineering, Lund Institute of Technology. Sweden.

Vejdirektoratet (1996), Miljöproriterede gennemfarter - effekter i 21 byer. (*Environmentally Adapted Through Roads — Effects in 21 Villages*). Rapport nr 70, 1996. Denmark.

VRU-TOO (1995), 'Final Report' VRU-TOO (*Vulnerable Road User Traffic Observation and Optimization*), DRIVE II Project V2005, ITS Working Paper 439, Leeds, UK.

Vägverket. (1997) Demonstrationsförsök med dynamisk hastighetsanpassning i tätort (*Field Trials of Dynamic Speed Adaptation Systems in Built-up Areas*). Publikation 1997:19, Borlänge. 1997. Sweden.

www.ingramcontent.com/pod-product-compliance
Lightning Source LLC
Chambersburg PA
CBHW081400170526
45166CB00010B/3154